Bees

Christine Butterworth

Silver Burdett Press, Morristown, New Jersey

Do you like honey?
Honey is made by bees.
They make it in a hive.
The hive is where the bees live.

2

The beekeeper takes care of the hives.
He puts them near flowers.
The bees need flowers
to make honey.
The bees store honey in
a honeycomb inside the hive.

The beekeeper has come
to get the honey.
He puffs smoke into the hive.
The smoke makes the bees sleepy.
Then he lifts the honeycomb out.
Look at the bees on this honeycomb!

4

Three kinds of bees live in the hive.

Most of the bees are small worker bees.

A few are a little bigger.

These are male bees. They are called drones.

The biggest bee is the queen.

In the picture you can see drones and
worker bees.

The worker bees do all the jobs
in the hive.
They make wax in their bodies.
They use the wax to make the honeycomb.
The honeycomb is made up
of little cells.

6

The worker bees fly out of the hive
to visit the flowers.
This bee has a long tongue.
The bee uses her tongue to drink
the sweet nectar from the flowers.

Can you see the yellow dust on the bee?
This is called pollen.
The worker bee has long hairs
on her back legs.
These hairs catch the pollen.
The bee brushes pollen onto
the hairs with her front legs.

The bees fly back to the hive.

They carry pollen on their legs.

They bring back the nectar they have drunk.

The pollen and nectar are food

for the bees in the hive.

Some worker bees guard the way
into the hive.
The guard bees use their feelers
to smell the new bees.
They smell the new bees to see
if they are friendly.

This wasp wants to get into the hive.
It has come to steal the honey.
Some guard bees sting it and kill it.
When the bees use their stingers
they die, too.

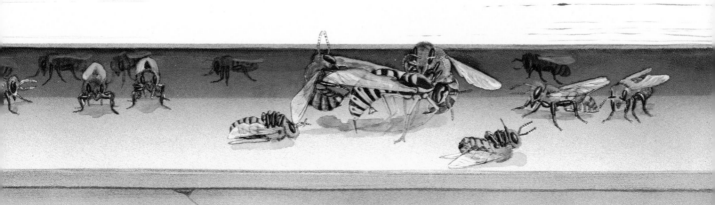

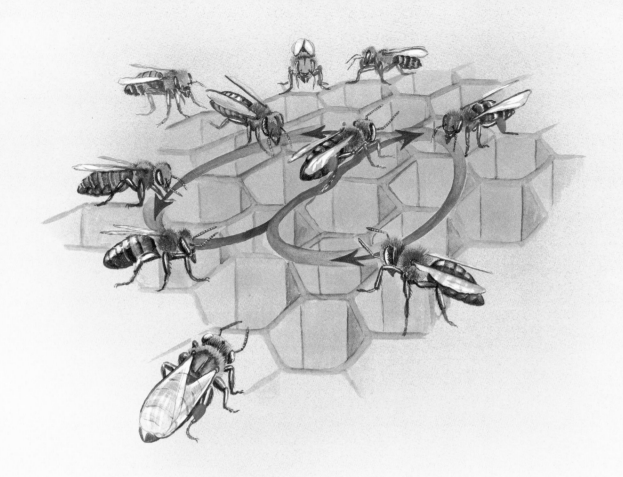

When the bees come back to the hive,
they do a dance.
They dance in a pattern.
This tells the other bees where
they can find flowers with nectar.

This bee is looking for an empty cell.
She will put the pollen from
her back legs into the cell.
Bees store pollen and honey
in the honeycomb.

The bees share the nectar they have
drunk with the other bees in the hive.
Each bee brings the nectar
back up into her mouth.
She puts the nectar on her long tongue.
She passes it to the other bees.
The nectar from the bee
turns into honey.

The worker bees put some honey
in the empty honeycomb cells.
Other worker bees make wax lids
for the cells.
The honey cells are a food supply
for the bees.
They will live on the honey
during the winter.

When the beekeeper takes
the honey, he gives the bees
sugar to eat instead.
This keeps them alive
during the winter, and
we still have honey to eat.

The bees stay inside the hive all winter.
They eat their food and stay
close together to keep warm.
In the spring they can leave the hive.
They fly out to find the flowers.
They collect pollen and nectar.

One day in the summer, a young
queen bee flies out of the hive.
The male drones fly after her.
They chase her as she flies.

18

The drones mate with
the queen as she flies.
After they mate, the drones die.
The queen bee goes back to the hive.

19

The queen stays in the hive.

She does not go out to find food.

She lays eggs all the time.

The worker bees feed her.

The queen looks for empty cells
in the honeycomb.
She lays an egg in each empty cell.
After three days the eggs hatch.
A tiny, white grub comes out
of each egg.

The grub is called a larva.
The worker bees feed the larva.
At first it eats rich dark food.
This is called bee milk.

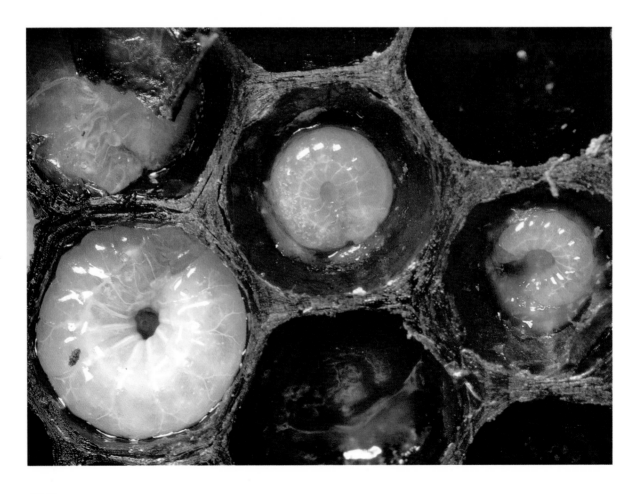

The larva grows very fast.
After a day it is five times bigger
than when it came out of the egg.
Now the worker bees feed it with honey.

When the larva is a week old,
it has no more room in the cell.
Now the larva stands on its tail.
Worker bees put a wax lid
on the cell.

Inside the cell the larva
puts silk all around its body.
The larva is now in a cocoon.
The larva will stay in the cocoon
until it grows into a bee.

The bee in the cocoon is big now.
It is time for it to come out of the cocoon.
The young bee bites around
the wax lid of the cell.
She pushes her way out.

When the bee comes out of the cell,
her skin is white and soft.
It takes a day for it to get
dark and hard.
Then she can start to do her work
in the hive.

The queen in this hive is old.

It is time to make new queens.

The worker bees make some big cells.

The old queen lays an egg
in each big cell.

The eggs hatch, and the grubs come out.

The worker bees feed each larva.
They give it rich royal jelly.
They do not give the larva
bee milk or honey.
The royal jelly makes each larva
turn into a queen bee.

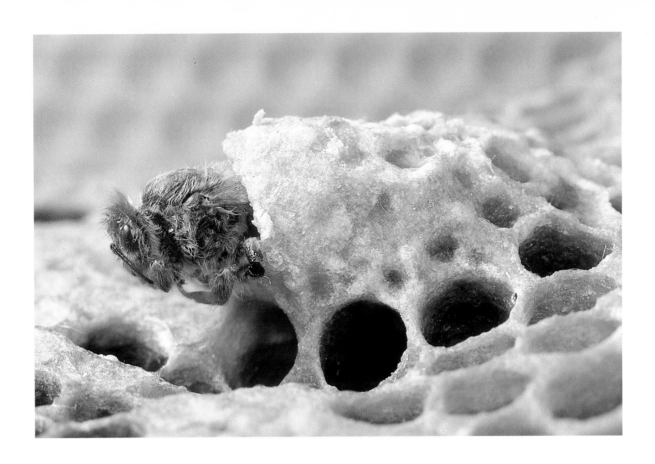

The young queen bees come
out of their cells.
The first queen to come out
kills all the rest.
A hive can only have one queen bee.

On the day the new queen comes out,
the old queen flies away.
Some of the worker bees go with her.
This crowd of bees is called a swarm.
The swarm looks for a new home.

This swarm has found a new home
in a tree trunk.
The worker bees fly out to collect
honey and pollen.
They start to build cells
to make a new honeycomb.

Reading consultant: Diana Bentley
Editorial consultant: Donna Bailey

Illustrated by Paula Chasty
Picture research by Suzanne Williams
Designed by Richard Garratt Design

First published in 1988 by
Macmillan Children's Books,
a division of Macmillan Publishers Limited
4 Little Essex Street, London WC2R 3LF and Basingstoke

Published in the United States by
Silver Burdett Press, Morristown, New Jersey.

Printed in Hong Kong

Library of Congress Cataloging-in-Publication Data
Butterworth, Christine.
  Bees.
  (My world)
  Summary: Discusses the behavior and characteristics
of honeybees and describes how they make honey.
  1. Honeybee——Juvenile literature.   [1. Honeybee.
2. Bees]   I. Chasty, Paula, ill.   II. Title.
III. Series: Butterworth, Christine.   My world.
QL568.A6B93 1988   595.79'9   87-23400
ISBN 0-382-09554-5

Photographs
Cover: OSF Picture Library/G. A. Maclean
Bruce Coleman: 4 and 16 (Jane Burton), 21, 23, 25, 28, 29 (all
   Prato), 32 (Jane Burton)
Frank Lane Picture Agency: 6 (Treat Davidson)
NHPA: 5, 13, 15, 20, 22, 30 (all Stephen Dalton)
OSF Picture Library: titlepage (David Thompson), 7 (J. A. L.
   Cooke), 9, 14, 17 and 24 (all David Thompson), 31 (Georgina
   Dew)
Zefa: 8

THIS BOOK BELONGS TO: